Beyond the Known

Horizon

Beyond the Known Horizon

S.B.Asoka Dissanayake

Asokaplus

Contents

Chapter 01

Introduction

Why another book?

All my previous books were not premeditated but imagination run riot. They had all the ingredients of human mind, pun, rant, satire, fiction, fascination, imagination, prejudices, predicaments and above all cross section of random thought moments.

Tragedy was taken out by default. I do not want to make readers more miserable than they already are. I gave little credence to scientific thinking but scientific observations were implanted amidst the material.

I went into an episode of digital zeroing for six weeks (Please read the note at the end of this book) and emerged by strange coincidence invigorated. The idea of this book was impregnated coming out of the halo of black hole in my mind.

In other words, creativity had a rare chance to enlighten the inner me. I had a fascination for physics and astronomy from my childhood but my vocation kept me out of it all my life.

This book is not about scientific fiction or friction.

This book is not on philosophy or theology.

It is not about abstract science too but something in between.

The black holes, dark matter and dark energy are the founding members or ingredients.

But talking about them in scientific terms is simply boring.

There is hardly any book on them but lot of junk out there.

I went in search of this material in the universe with my imagination run riot.

I had the company of beautiful collection of photographs taken by the Hubble Telescope.

Looking at those pictures was not like looking at dark X-rays.

Well, well, that made me to postulate a new hypothesis in theory not at all, tested scientifically.

The dark matter is so docile one cannot take a grain of it weigh it and test it in the laboratory.

Neither, you could taste it nor smell it but it is out there waiting to shake hand with you, outmatching the matter by 1 is to 10 ratio or even more.

Writing about it will stir up controversy.

That is not my intention, though.

The laboratory is out there to discover.

I delve into, not so deep and this book is the outcome.

To make it interesting from the very beginning I take the "Big Bang" to misery so it folds up and goes into hiding or a "Big Crunch".

If I don't do that nobody will read this book.

Only a little is written about the dark mater to make it less boring for the reader.

The reading material is in a twisted form of a string.

The string theory of matter in action.

I prefer the string theory here (not my conviction anyway) since it gives so many dimensions to matter real.

The core material the dark matter and dark force are tightly wound up with the matter real.

One has to take the strings of material luminance out of the tight rope, string by string and picture it as at was, a halo of discovery.

Don't hold onto the rope.

Take the strings out and discover the dark secrets.

Luck is on you and you will discover more if you also let the scientific brain of you released from the matter real of the visible world.

I hope you would not end up mad after reading this book.

The sole purpose of this book is the enlightenment and bring the sanity out of insanity of the material science.

Dark matter (dark force included) is in fact a scientific saint!

Chapter 02

Looking through the Tinted Glasses

Talking a bit about the known mater is not an anticlimax as a prelude, before delving into the unknown horizon.

We are obsessed with visible matter and gravity.

Without gravity we cannot exist.

Our lungs will not inflate but rather collapse in a vacuum and our bones will crack.

Equally, we won't be able to stand up in a sphere of 3 or 4 Gravitational Force.

The gravity alone is not enough.

We need our vision to orientate in space and see the matter embedded in space.

If our species was born totally blind how far we would have progressed in evolution?

That is a point to ponder.

Ability to see and use energy spectrum is common to both plant and animal kingdoms.

But this obsession with light and the wake and sleep cycle have blinded our brains to ignore the unknown.

In fact, we fear darkness.

Imagine there were no stars in the sky at night and total darkness only prevails.

We would fear the sky and never bother to look at.

We were so obsessed with matter or brightness our ancestors described the sky as heaven.

They even created a mighty god supervising these heavenly stars.

But with the development of the telescope and later the Hubble telescope our looking at the matter intensified and did not retard or wane.

What it actually retarded was, the looking at the so called empty space of no real value until of course Hubble discovered the expanding universe.

The expanding space.

In fact, the matter is finite or may be exhausting itself by constant burn or transformation but the space is expanding..

The matter currently at less than 5% given the time of billion light years will disappear to subatomic particles.

One does not need a "Big Bang" or "Big Crunch" to foresee this eventuality.

The application of the current physical dimensions will not fit the bill as it were, with the emergence of dark matter into scientific thinking and existence.

An alternative science is mandatory and dark matter has provided this by a stroke of luck.

We also need to get out of the mentality of the last 100 years that we need to test every possibility with equations and perform tests to verify our understanding.

The dark matter and dark forces would not obey current physics. But it would appear in relative terms or relativity since it is finely tuned with the matter but without reacting.

It only does the warping.

One should not get confused with big "Black Holes".

Black Holes are not where dark matter is concentrated even though, it could possibly have some within a black hole, giving it much needed transparency (then the light would travel through the pit and come out on the other end and give us a faint glimpse of hope of visualizing the bottomless pit).

It is a phenomenon where light gets pulled by the gravity of the big hole and won't let us see the pit. What is within a black hole is unknown, just like that we know very little of the physical properties of the dark matter.

Dark matter is different to black holes..

It is out there in plenty in the galaxies.

It is like a fish tank with water and no fish or plants in it.

So we should describe the dark matter as a totally transparent tank (not visible to our spectrum of energy) full of matter but we cannot grasp or fathom it with the physical spectrum of energy we can manipulate.

We cannot put matter like fish and plants into this empty tank to see its relation to relativity of the known physics.

Certainly, we cannot do that on this planet earth. This planet is full of matter with no substantial quantity of dark matter.

If dark mater is there, they are made more invisible or integrated with matter real, as a form of some nuclear forces yet undiscovered.

I believe the dark mater, if they are actually reacting with the matter they may actually contribute to the mass where modern particle physics find it difficult to accurately measure the mass in real quantities (in a known quantity of atoms).

That is the beauty of the dark matter.

They do not obey physics that we know of.

That means we give up and only postulate its behavior.

No, No, that is not the way.

If we look up to the sky, all the experiments are going on, up in the heaven of galaxies.

In my belief, the geometry of our galaxies whether they are exploding or collapsing, the real force that tilts the balance is the dark force and dark matter.

They govern the very existence of matter real.

That is my outrageous hypothesis which I want to elaborate a bit not in abstruse terms or complicated equations but with simple common sense.

Let somebody else do the calculation and discover the alternative physics behind them.

Chapter 03

Why I disagree with the "Big Bang"?

Below is 'quote I picked up from Pinterest.

It is a good site to visit when one is lost for a thought moment.

My current analysis should not fall into any of the three IDEAS, EVENTS or PEOPLE.

I wonder where it should fall BUT the reader should not make any cross reference or opinions, lest s/he falls into the TRAP set in by the Big Bang itself.

Great minds discuss ideas, average minds, events, small minds, people, that was my pick from the Pinterest.

My reasons for not believing the Big Bang are;

1. Bing Bang sounds nice as a phrase but it falls short of Reality of Nature.

2. That is of course if there is something called Reality, which the human mind can grasp at will.

3. The bottom line is we cannot even grasp what is called Mind and its physical constituents.

4. The Mind is so labile by the time it is measured by a physical agent (for example an EEG) it has changed into another form of expression.

5. It begs the question can something that is so labile can measure something so enormous like the universe in its full motion.

6. Mind has to FREEZE the universe out of its motion for it to be measured in physical terms.

7. It does not mean physics and derived equations can explain everything.

8. There are limits and boundaries for physical laws and the bone of contention here is does the Universe end where physical laws fails to apply, for example near the horizon of a "Black Hole.

Let me express my discontent about the Big Bang.

Number one is, it freezes the universe at Time Zero or 15 billion light years ago and then work forward from there.

Why is it not 50 billion light years?

Time is a concept related to motion of particles (for example electrons).

It is not a constant but a certain relationship to what we call SPACE.

If that is so the physicist should be able to send a particle equivalent of an electron to the universe and wait for another 30 billion years for it to report back to us of the outer limit.

But that electron or particle has to be guided from the time of its exit from the emitter through SPACE and should not be subjected to any interference of any kind till it reaches the boundary of the universe.

And then WHAT?

Would it turn BACK?

Would it escape the laws of physics?

This argument can be stretched to any limit till cows come home but would not answer the strangeness of the whole universe.

That is the experiment side of it which is not worth performing since our life cycle is limited by about 100 years, insignificant in light years.

However, one should not bother about experiments, the mother nature is performing experiments at will and they are abound.

This is what Hubble showed us with the spectrum of light which the human eye can see, the Red shift equivalent to receding universe and Blue shift as of advancing objects.

There is another assumption that is light or the photon moves at a constant speed of light.

No deflection, reflection or rarefaction is allowed in this concept.

It travels head on.

Why?

How can something be so constant in a universe where everything else is changing including the SPACE itself.

If time is a relative concept, the speed of light is also a concept by itself.

This is the bottom line in relativity and in theoretical physics.

If we don't have a constant and a time frame we cannot have objective measurement of anything in space.

In other words FREEZING the Universe at time ZERO suppose to explain, everything including the universal truth.

Now I come to the crunch point which is "How did the Mass at Zero Time Originate"?

There is no answer to this point and it is in the same line of framing as, the mighty CREATOR was at work and did have a say.

In other words Big Bang was created to fall in line with Holy One.

That was the biggest fallacy.

Currently particle physicists are struggling to explain from where the Mass of an atom comes?

They are beginning to realize that even the space is not empty but has virtual particles, so that the other particles can make stable environments or tiny universes.

The presumption is that the space is expanding not the particles that are moving at variable constants in relation to the mysterious gravitational forces making us believe that the particles are moving and not the other way round.

That is again the beauty of relativity.

So my unifying argument (mind you a theory) is that there are tiny universes within this universe and there are many tiny universes outside of this universe.

The words 'tiny' and 'this' are also relative terms not absolutes.

There is lot in physics left to be discovered and that is why Steven Hawking talked about Many Worlds and the bosons or the God Particle (no reference is made to the god) to describe the unknown force.

The Spectrum is not limited by what is visible, it stretches beyond and outside the visible spectrum.

Mind you the grand theory is yet to be discovered.

I want the young physicists to take the mantle and explore the nature within and outside our universe.

Chapter 04

There are two Big Holes in the Big Bang Theory

How come something comes from nothing?

It is akin to creation by god.

Unfortunately, scientists a century ago had to satisfy the Church before uttering any theories fearing misgiving by the Church.

The Church was the authority and it had to ratify.

It was a pity.

Second was the inadequate explanation of the dark matter, a constant was proposed but this did no go well with the contemporaries.

Why there was so much dark mater was never adequately explained.

The possibility of a "Big Crunch" as opposed "Big Bang" was excluded by default.

There is lot of dark matter amidst our galaxy.

We know the matter is consumed in the process of emitting photons and slowly but surely matter is exhausting itself.

There is transformation not annihilation.

Is it possible that the dark mater embedded in our galaxy is a byproduct of this transformation?

If the dark matter is expanding we got to seriously consider this probability.

If the dark mater is decreasing (in any galaxy) my theory has to be discarded outright.

If dark matter is stable we got to find an appropriate explanation.

How it got trapped amidst the matter of our galaxy (other galaxies, too) need some explanation.

Black holes of dying stars may be a candidate BUT black holes by themselves are a mystery to modern physics.

Chapter 05

This is my alternative theory to counteract the "Big Bang"

I have said, I am uncomfortable with the "Big Bang" and if I do not propose an alternative theory, then my scientific reasoning becomes VOID.

I want to propose a random theory not a unifying one.

It may sound bizarre but it is tangible in theory.

The space is taken into account with its virtual particles.

In this theory of "Whole of the Universe" is taken as one conglomerate.

The matter real is only the small change.

It is less than 5%.

The dark matter is said to be 25%.

The rest 70% (near enough) is dark energy.

My assumption is that matter which carry energy with charge would always move towards a more stable state without charge.

Only way for it to make that move is to become dark matter and in that process it releases dark energy and dark particles without charge.

This in fact, makes the space expand.

The matter moving out and universe expanding are two sides of the coin.

It is the virtual reality.

In this theory the matter reduces not by constant shift but by random events.

So when the matter is reduced to a critical point, say less than 1% something intangible has to happen.

There is a mismatch.

At this point Dark energy and Dark Matter cannot rule the universe.

It has to go into reverse gear, not in a "Big Bang" but in a series of steps.

So the dark energy and dark matter is consumed and new matter is formed and the worlds, galaxies and new universes are formed.

There is neither beginning nor end, only a process, somewhat akin to cyclical phenomena (forward-dark matter and backwards-transformation to matter) are in existence.

Better way to visualize is zipping and unzipping of a fastener.

The term cyclical is a misnomer for this transformation of energy particles.

It falls into quantum dynamics, not a cycle, like day and night.

It is not like birth or bust or bubble.

There is only two or three dimensions to this theory but if few more dimensions are mathematically extrapolated, the possibilities are infinite.

Dark matter is docile in its very nature.

It tend to keep company with likes of it.

Since it has no electromagnetic forces or charges within its particles, there is no wonder they keep company.

What is keeping them intact without wandering away?

So we have to postulate there is dark force keeping them together in its own gravitation sphere.

This gravitational force is different from the gravitational forces binding matter.

In fact, it has anti-gravitational power and makes the universe expand.

In this scenario space expands and the matter real (like a pictures drawn on the surface of a balloon) all around, gets pulled apart together.

Even though, it portent to be docile, it is not.

It makes a real contribution to the existence of the matter.

I go even further and postulate that dark matter changes to matter given the right circumstances.

That of course is the core material in this book.

Suppose a bit of dark material in its solo shell come into the vicinity of a galaxy which has enough and more dark matter to pull them towards.

What is the likely outcome.

Simple explanation is for it to unite with the likes.

Nothing else but expanding space within the galaxy..

The other scenario is to strip the galaxy of its dark mater.

The galaxy has to contract within and the dark matter finds its own space in the universe.

But what if this process is different altogether, simply because of its docile nature.

The two dark bodies would come into near enough contact range but without merging but wandering about happily.

What tilts the balance is not dark matter but matter itself.

For instance, the material in a star is burning intensely and spinning violently.

By random event, it strips a string out of the dark matter on its outer periphery,from the solo unit..

The dark mater spins into a different mode by this action.

It starts forming matter and start striping little bit more from the dark matter.

What starts as a random event now blows out of proportion to become a hydrogen star.

A star is born.

Not one may be more!

The simplest of atom, the hydrogen is the outcome when matter is born out of dark matter transformation.

The way to test this theory is to test whether matter can be changed to dark matter or dark forces.

This is more feasible since only charged particles are necessary to neutralize the charged particles of matter and certainly not antimatter.

No annihilation is envisaged.

The converse of it is to try change dark matter to matter.

Bombarding dark mater to make matter is assumed more difficult due to its stability and its inherent expansion.

In other words contracting (compacting or concentrating space) the space is insurmountable.

If that can be achieved "the time travel" actually becomes feasible.

This is my alternative theory to counteract the "Big Bang".

Chapter 06

Dark Matter is the Root of the Matter

My theory of Matter verses Dark Matter or their transformation has many applications.

This time it is related to the galaxies.

How can one explains the geometry of galaxies.

There are many;

Spiral

Globular

Flat Disk

Or any combinations the above.

Since most of the galaxies are either pulling apart or pulling together towards each other, anyone of these formations are possible.

It is the dark matter that gives the bulk of gravitational pull and the shape not the burning and changing stars.

The contribution of the stars are relative since dark matter far exceeds the matter in real terms.

Our Milky Way is the best example.

The dark matter and its gravitational pull gives the right balance to the Milky Way.

Even though, it is receding it does not become fully blown away.

Similarly it does not pull together and goes into a "Big Crunch or collapsing galaxy / universe.

The disk shape of our universe has something to do with this behavior and maintaining the correct balance.

Subtle variation of the position of the stars counter balances the pull (matter) and push (Hubble constant) of our Milky Way.

Even though, it expands the reorientation of the stars (matter) and dark matter gives the unusual stability to our Milky Way.

The intention of this piece is to expand the implications or the working of the theory proposed.

In spiral galaxies the outer rings or tentacles are moving faster than the inner stars.

Strangely though, outer masses are associated with dark matter.

Hold on.

This is probably where the matter is changing into dark matter.

It is unlike the bombardment described in particle physics experiments.

The speed at which these galaxies are moving in tandem with dark matter gives the opportunity for realignment of subatomic particles.

Matter at the margin turns into dark matter by random events to begin with.

Its gives the stability and this pushes it further on to make more dark matter with gravitational pull.

Let us assume ultimately all the matter is transformed into dark matter in a particular point in space and form into strings twisted tight like the DNA (string theory of many dimensions) with enormous gravitational pull.

So what happens next?

Invariably by random event it gets close to a galaxy with right amount of dark matter and bit more to overcome the gravitational pull of the unitary or solitary dark matter without stars already in its sphere of influence.

The galaxy with dark matter stripes a string out of the solo dark matter.

Behold!

This slightest strip and imbalance it creates, the solo dark matter turns table and starts becoming matter again.

The theory to counter the "Big Bang" is born real.

The globular and elliptical galaxies have their own ways.

I will work out their mechanics in subsequent chapters.

The current attempt is to show the feasibility of the theory sparkling random events at galaxy scale.

Birth, Boom, and Bust are not the intentions of the overall discussion.

Is it possible to gather some tangible views and improve on this theory or the hypothesis?

Chapter 07

The Creativity of Darkness

Darkness has connotation of evil.

If not for the darkness we miss the cooling and restful periods at night. Once a day spin of our planet to head away from the sun and direct us to the night sky with shining stars was the beginning of astronomy.

The overall construction of the heaven would not have been possible, if the sun was shining 24 hours a day.

The darkness was the blessing for the scientist.

Not only darkness made him to invent the light bulb but the telescope too.

If we did not have a day and night cycle, the science of astronomy would have never come into being.

Eternal days with strong radiation would have made life impossible. The water would evaporate and the atmosphere would deplete itself of oxygen after full combustion of matter on the earth.

In any case oxygen would have never formed in the earth atmosphere.

When all the rest is sleeping only a few astronomers would gaze the sky and discover pulsars, quasars, exploding suns and new galaxies.

That was how the pioneering astronomers worked to built on Copernicus and Galileo.

Now most of these are done automatically by the computers and the photographic data generated are fed into supercomputers to produce simulations.

Even the super computers depend on the data gathered at night.

They are useless come day light for gathering information.

The humans are obsessed with luminance or sunlight or artificial lights.

In cities the telescopes cannot recode accurate data due to interference from human sources of light.

That is why most of the telescopes are located in remote places well away from human habitations.

Chapter 08

What are the Applications of Dark Force?

What are the Applications of Dark Force?

This question is the crunch point itself.

We are obsessed with travel through the space but what we can achieve is very little in terms of the limitless universe.

If we assume there are life forms out there, we need to communicate with them or if possible visit them in real time.

That is my imagination run riot.

Alternative physics is the only way to change the current predicament (inability to reach).

There should be a way to counteract gravity.

The dark force has anti-gravity built in.

Unlike matter when it reacts with antimatter that annihilates, it does not release giant energy to annihilate everything. It subtly merge with the galaxies and make matter worth illuminating (gives the contrast to the visible matter) and looking at.

We need to get into its core of gravity in a capsule made of virtual dark matter and built a time tunnel to travel through space forward, keeping our body made of matter intact and without aging (problem of aging solved for a brief moment of universal time), if not backwards in time.

I prefer the backward travel then I can gain some credibility to destroy the Big Bang" theory, once and for all.

This is an impossible scenario to think or dream about.

But if we can go beyond the current physics and find a way to utilize dark matter out there in the universe for our navigation in space, the time travel is the next dimension of our thinking.

The dark matter can warp time and space.

In other words contracting the space would mean we are advancing towards a nearby galaxy (blue shift instead red shift).

The universe becomes a football field in this dimension.

The new physics is the ball in play!

Chapter 09

Types of Galaxies

Elliptical galaxies

Elliptical galaxies are the most abundant type of galaxies found in the universe. They make up to 60% of all the galaxies. They are the oldest.

However, because of their age and dim qualities, they are frequently outshone by younger, brighter collection of stars.

Elliptical galaxies lack the swirling arms of their more well known cousins, the spiral galaxies.

Instead, they bear the rounded shape of an ellipse.

Spiral galaxies

Spiral galaxies take their name from the winding spiral shape they demonstrate.

30 percent of the galaxies in the universe observed by scientists are spiral galaxies.

These twisted collections of stars and gas often have beautiful shapes and are made up of hot young stars.

Most spiral galaxies contain a central bulge surrounded by a flat rotating disk of stars. Made up of older, dimmer stars, the bulge in the center is thought to contain a massive black hole.

The dim light from the older stars can make the bulge difficult to pinpoint, and there are some spirals that lack this characteristic.

Starburst or Irregular Galaxies

They do not fit into either category described above. About 10 percent are irregular galaxies.

They are probably the youngest galaxies as shown by intense activity and new star formation.

Astronomers have found evidence of disk of gas between the stars and they are the source of new star formation.

Galaxy Merger

The galaxies merge with each other due to their massive gravitational forces. They are referred to as the car crash victims of the universe.

The resulting galaxy merger is violent and dramatic and change the structure and shape of the clashing galaxies.

Dark matter within galaxies

There is dark matter amidst the stars in these galaxies and the exact amount of it is not estimated. Now is the time for me to expand on my hypothesis further in reference to globular or elliptical galaxies and Starburst galaxies.

In my opinion dark mater does not sit idle.

Apart from giving the geometrical shape and stability, they are in the process of transformation.

The spiral galaxies have been used by me to explain the dark mater becoming matter by random events.

It is the matter of pulling apart a string of dark matter from a saturated or solo dark matter hovering at the periphery of a galaxy containing right amount of the dark matter, probably slightly in excess of the wondering solo dark matter, that trigger the formation of matter real as opposed to invisible dark matter.

This triggering event is similar to Big Bang but in a smaller scale and dimension (mind you this is my alternative theory to counteract the Big Bang).

The dark matter and its force goes into reverse gear and start producing matter in series of step quite invincible to the observer. It just mergers into the background of the galaxy without a big bang.

The matter that is formed has to be the simplest of all.

That is hydrogen atom.

The dark force that is released drives the reaction of fusion of hydrogen to form helium in these young stars.

New stars forming in the spirals which are blue in color is the end result.

Here the gravitational forces of the matter (the galaxy) is in fine tune with the dark forces and dark matter.

There are no shooting or exploding stars but it is a maternity home for new stars in formation and trying to defy the inbuilt gravitational forces by making spiral formations.

But they cannot defy the gravitational force of the newly formed stars and the gravity within the dark matter.

In fact, the stability is given by the dark matter itself.

The dark matter occupying the space let this happen gently and it gives the property of expansion in space to operate in tandem with new star formation.

In this scenario, Starburst formation (there is plenty of gas burning into stars in Starburst formation and its appearance is the result of these gases burning) is the most likely at an earlier stage of formation where plenty of burning gas is available, giving rise to spiral formation at a later stage and still later when the gravitational forces far exceeds the expansion allowed by dark forces, elliptical galaxies form.

In fact elliptical galaxies are the antithesis to dark matter transforming to matter.

Deep in the globular galaxies matter transforms into another dimension by forming black holes by shear effect of the gravity. The globular galaxies may have traces of dark mater in its periphery but nothing substantial deep in its center.

Its center is occupied by black hole which are invisible both due to warping of time and space dimensions and the interference of the light emanating from the globular cluster of stars in the outer perimeter of the globular formation.

So my theory in evolution has a place for dark holes too.

In my theory there is no divine intervention but matter and dark matter in transformation not in chaos but in random fashion.

Matter and dark matter are two sides of the coin with opposing properties.

One is visible the other is invisible.

Our obsession to matter visible has made us to ignore something as substantial, may be 10 to 100 times more than the matter real.

Our obsession has a blinding effect.

It is a continuum like, the physical properties of energy we measure in physics but in a different dimension merging imperceptibly in all dimensions as shown in the formation of galaxies.

Its wonder is, it needs different type of quantum physics to grasp it.

I call it "the humble pie effect" of not knowing the alternative physics.

My bone of contention is that the Big Bang has no explanation for the enormity of the dark matter and dark holes.

Black Holes

Black holes are some of the strangest and most fascinating objects found in outer space.

They are objects of extreme density, with such strong gravitational attraction that even light cannot escape from their grasp if it comes near enough.

Albert Einstein first predicted black holes in 1916 with his general theory of relativity.

The term "black hole" was coined in 1967 by American astronomer John Wheeler, and the first one was discovered in 1971.

Chapter 10

What Is Dark Energy?

In the early explanations nothing was fairly certain about the expansion of the Universe.

It was thought it might have enough energy density to stop its expansion. The converse was also thought to be right. It had enough energy to counteract the force of gravity preventing total collapse.

That is the view of the stable universe.

The other point of view was it might have so little energy density that it would never stop expanding.

The unstable state of the universe.

It was supposed that the gravity would slow the expansion, as time went on.

It is in fact is expanding!

However, the slowing had not been observed, but, theoretically, the universe had to slow. The universe is full of matter and the attractive force of gravity pulls all matter together.

This was the view before dark matter was discovered.

Then came 1998 and the Hubble Space Telescope (HST) observations of very distant supernovae that showed that, a long time ago, the universe was actually expanding more slowly than it is today.

So the expansion of the universe has not been slowing due to gravity, as everyone thought, it has been accelerating.

No one expected this, no one knew how to explain it.

But something was causing it.

Eventually theorists came up with several explanations. Maybe it was a result of a long discarded version of Einstein's theory of gravity, one that contained what was called a "cosmological constant."

Maybe there was some strange kind of energy that filled the space. Maybe there is something wrong with Einstein's theory of gravity and a new theory could include some kind of field that creates this cosmic acceleration.

Theorists still don't know what the correct explanation is, but they have given the solution a name.

It is called the dark energy.

Dark Energy and Expanding Universe

With the big bang theory it was proposed that the universe is expanding. The rate of expansion since birth of the universe 15 billion years ago was not constant.

It was slow at the beginning.

It started expanding faster at about 7.5 billion years ago, when objects in the universe began flying apart at a faster rate. Astronomers theorize that the faster expansion rate is due to a mysterious, dark force that is pulling galaxies apart.

More is unknown than is known about the dark matter.

We know how much dark energy there is because we know how it affects the universe's expansion.

Other than that, it is a complete mystery.

But it is an important mystery.

It turns out that roughly 68% of the universe is dark energy.

Dark matter makes up about 27%.

The rest - everything on Earth, everything ever observed with all of our instruments, all normal matter - adds up to less than 5% of the universe.

Come to think of it, maybe it shouldn't be called "normal" matter at all, since it is such a small fraction of the universe.

One explanation for dark energy is that it is a property of space. Albert Einstein was the first person to realize that empty space is not nothing.

Space has amazing properties, many of which are just beginning to be understood.

The first property that Einstein discovered is that it is possible for more space to come into existence.

The two biggest mysteries in cosmology may be one.

A new theory says that dark matter and dark energy could arise from a single dark fluid that permeates the whole universe.

What this could mean for earth based dark matter search is that we do not have the capacity to reach and interact with dark matter.

Dark matter, as originally hypothesized, is extra hidden mass that astrophysicists calculate is necessary for holding together fast moving galaxies.

The most popular notion is that this matter is made of some yet to be identified particle that has almost no interactions with light or ordinary matter.

Yet it seems to be everywhere, acting as a scaffolding for galaxy clusters and the whole structure of the universe.

On the other hand, dark energy is needed to explain the more recently discovered acceleration of the universe's expansion. It supposedly exists all throughout space counteracting with gravity.

After decades of studying dark matter scientists have repeatedly found evidence of what it cannot be but very few signs of what it is.

That might have just changed.

A study of four colliding galaxies for the first time suggests that the dark matter in them may be interacting with itself through some unknown force other than gravity that has no effect on ordinary matter. The finding could be a significant clue as to what comprises the invisible stuff that is thought to contribute 25 percent of the universe.

The simplest model of dark matter portrays it as a single particle one that happens to interact with others of its kind and little or no reaction at all with the visible matter.

For dark matter to interact with itself requires not only dark matter particles but also a dark force to govern their interactions and dark boson particles to carry this force.

This more complex picture mirrors our understanding of normal matter particles, which interact through force carrying particles. For example, protons interact through the electromagnetic force, which is carried by particles called photons (particles of light).

Now scientists led by Richard Massey at Durham University in England report in Monthly Notices of the Royal Astronomical Society the first signs that dark forces and dark bosons might really exist.

Researchers used the MUSE (Multi Unit Spectroscopic Explorer) instrument, the very large telescope in Chile, along with the Hubble Space Telescope to examine the Abell 3827 cluster, where four galaxies are colliding in a cosmic car wreck.

To determine where the invisible dark matter lies, astronomers took advantage of a natural phenomenon called gravitational lensing, predicted by Einstein's general theory of relativity.

Lensing occurs when mass warps space-time, causing light traveling across a dark region with dark matter to take a curved path. The dark matter in Abell 3827 is plentiful, so it warps the space around it significantly.

When light from a distant object behind the cluster travels to earth, it passes through this distorted area and produces telltale signs of lensing, such as arcs of light and double images, that astronomers used to "weigh" the unseen matter in the cluster.

The scientists found that in at least one of the colliding galaxies the dark matter in the galaxy had become separated from its stars and other visible matter by about 5,000 light-years.

One explanation is that the dark matter from this galaxy interacted with dark matter from one of the other galaxies flying by it, and these interactions slowed it down, causing it to separate and lag behind the normal matter.

The interactions would be similar to what happens when two protons pass near one another. Each releases a photon that is absorbed by the other, causing both particles to recoil. This repellent force happens between any two particles with the same electromagnetic charge and it could happen between any two dark matter particles as well.

But because dark matter is not affected by the electromagnetic force, only a new "dark" force, carried by a so called dark photon, could produce the repulsion.

It could also be that only a portion of the dark matter interacts with itself whereas the bulk of it is a more traditional single particle type.

It does seem, this is the kind of model that fits and explains, the observed behavior, in which only a small fraction of the dark matter interacts, says Harvard University physicist Lisa Randall, who has envisioned such a model.

I tend to agree with this model and in my hypothesis only a fraction of the dark matter reacts in the beginning and changes to matter real. This initial change causes a mismatch of normally docile dark matter to turn table and form matter that may end up forming young stars.

I am against the perfect symmetry of dark matter or string hypothesis of many dimension for dark matter.

I tend to vision it like a zip fastener, but with a difference. The zip fastener is zipped in many points and unzipped in the ends.

It keeps zipping and unzipping.

Zipping gives it a virtual stability.
Binds the galaxies together..

Unzipping gives the expansion.
Unzipping gives it many dimensions including matter real and dark force and expansion of space.

The zipping model can produce to spiral and globular galaxies.

Only laboratory we have to test these evens is out there in the sky. One has to observe the collapsing or colliding galaxies to formulate the mechanics behind them.

Chapter 11

Gravitational Lensing

Lensing has been used to help verify the existence of dark matter itself. The Bullet Cluster has been observed in both optical (visible) light and in X-ray. The majority of the light coming from the Bullet cluster comes from hot X-ray emitting gas, and has been overlaid onto the visible light image. Superimposed on visible light and X-rays is the lensing effect of dark matter which can be measured by the lensing effect of the visible light.

The visible light of the galaxies helps to determine the lensing effect and its density.

During the collision, the baryonic X-ray gas particles (the 'normal' matter) will interact with each other through both gravity and electrostatic forces, slowing and shocking one another.

The dark matter particles, however, only interact through gravity and can pass through each other unimpeded by electrostatic interactions.

This means that the X-ray gas lags behind the dark matter as the two clusters escape the collision, causing the observed offset - most of the visible matter should be in the center of the collusion but lensing effect puts most of the mass further out.

Some scientists believe since the only observable effect of dark matter is gravitational, then perhaps our understanding of gravity is incomplete. It is possible that we are not observing a new type of matter, but that the laws of gravity as we understand them are wrong.

As a result, many different modified gravity theories have arisen to explain the dark matter phenomenon.

The Bullet cluster provides strong evidence for the existence of dark matter, as this offset between the light and mass is exactly what scientists expect to see if dark matter is real.

If we know something about the distances to the galaxies we look at with our telescopes, lensing can also tell us about the nature of dark energy because the amount of dark energy affects how galaxies and clusters form and develop. By using the distance between dark matter and the observer and the degree of gravitational lensing it causes when observing, the observer (telescope) would help us to measure the amount of dark energy in the universe to a higher degree of precision.

Then computer modeling with correct equations to simulate dark energy would give us the picture of the universe which we cannot reach by physical means.

The light from distant galaxies began traveling towards us many millions (or even billions) of years ago, providing a window into the early universe.

This means that it is also possible to work out if the amount of dark energy changes over time by observing galaxy structures at different distances from us.

Thus, gravitational lensing is a clean probe of the universe and has much to tell us about its two most mysterious components - dark matter and dark energy.

In fact, this is one way in which gravitational lensing differs from optical lensing, as gravitational lensing is independent of the wavelength (color) of the light. All light rays are bent the same amount by gravity. Optical lenses cause light of different colors to bend by varying amounts in a process called diffraction, resulting in the splitting of light into rainbows.

There is no such analogous effect with gravitational lensing.

Chapter 12

Does Everything carry or has Energy?

This sounds a silly question in modern physics.

Everything that has mass carry energy.

That is what happen when gasoline burns.

All matter can be burned to produce energy but the conversion strategy is difficult when the mass concentration is heavy.

Say for uranium.

Spontaneous decay of heavy metals results in cosmic radiation and we do not have to burn our fingers, try triggering mass conversion of heavy metal to energy.

They are unstable by nature.

When in motion this energy become exponential till the speed of light is reached.

In summary there are three forces of energy.

What is universal is gravity related to the mass.

Then there is electromagnetic force which needs a magnetic field and moving particle, such as electrons.

The third is cosmic radiation that comes from burning stars. This energy that comes from atomic fusion or fission are divided into weak and strong nuclear forces to fit in with the particle physics.

But what about the dark energy which has no known counterpart seen in matter.

Is it only a postulate?

Not real!

Gravitational lensing has taken this doubt out of the equation.

It is an energy that tries to counteract gravity.

The expansion of the universe has given credence to this premise.

But can we measure it?

As of now we can see it by indirect methods such as gravitational lensing.

Unlike the lensing that occur with light (photons), there is only one path for the lensing to manifest not multiples.

So objective measurement will defy physics as is known today.

It is the terminology that we are trapped in.

That is the very reason in my arguments, matter and dark matter are taken as two sides of the coin not as a mirror image.

Mirror image is OK for matter and antimatter but not for dark matter.

It is transparent to matter and lets the matter coexists in the virtual space it has created for matter.

It does not defy relativity but it merges amicably with relativity.

That is why I say and argue alternative physics is wanting.

I go further and reiterate dark matter is essential for the very existence of matter real.

For want of an analogy let us say, it does not have energy forces like in matter BUT has quantum of relativity (activity) to match every bit of energy the matter present in the universe.

In fact, it has more of it than matter and that is why galaxies form or galaxies collide and the universe is expanding.

Authors Note

Zeroing Digitally

Even though, I proposed Digital Zeroing more than two years ago, I could not practice it for a significant length of time.

I managed it for six weeks, this time round and the outcome was more than a PAID Meditation Session.

This book is the end result of digital zeroing.

It takes you to the creative plane of your mind, which is generally hidden behind the mundane foreground activity.

If one is short of ideas discharging already highly charged brain is mandatory.

It is like the dark energy described in this book.

Discover your brains dark energy by utilizing this technique, once every five years or so.

Do not try this on yearly basis and it won't work.

One has to let the dark energy accumulate in smaller doses.

Sleep takes one's working hours hyper charge away but the brain's inherent property is to keep some of the dark but creative energy entangled in the memory network unable to disentangles at will.

This is not the only way to discover one's hidden potentials.

There aught to be many more avenues.

It was invigorating and I ended up investigating bare bone physics.

Quantum Dynamics and space travel were few ingredients.

I must tell you it is more difficult than practicing simple Moment Meditation.

I had the right preparation over the last decade.

1. TV and Media I gave up many moons ago.

Not because I was preparing for digital zeroing BUT simply because the contents were absolute muck.

2. Radio was even worse.

I could manage the irritating mosquitoes.

I have the option of killing them but not the Radio Stars.

3. The digital media were so out of date, it is not worth visiting them.

I had given up reading even Sunday Paper, Sri-Lanka cricket included.

Only time, I spent little time was before the election.

4. Internet and email was the most difficult for zeroing.

My regular blogging almost one piece a day like a editorials in a newspaper was the biggest addiction.

Even though I dismantled 10 computers, I did buy a 64 bit new computer.

5. It was made easy with the new tablets coming into existence, killing my Linux blogging.

6. Digital cell phone (land base telephone is now in rare use) was never my companion.

I think it is the most destructive (for intellectual activity) piece of innovation.

I dare not state it is not a useful gadget, especially as a traveling companion.

7. One need to be out of work or in retirement for one to practice digital zeroing.

8. The most important prerequisite is the ability to SLEEP (even traveling standing in a bus) at the drop of a hat at any time of the day.

9. Except eating, drinking and going to the toilet one should not have a daily routine.

10. Getting rid of the calender is also important.

The idea is to forget what time of the day it or what is today.

With all the ingredients ready, I plunged into the unknown.

It was difficult for the first week BUT with perseverance, I got into the mode of enjoying doing the least bit of work.

With this change, even feeding the dog was automatic response.

Feeding the fish (I must say it was a big mistake to start keeping fish with temperature soring to 35 degrees outdoors and they die above 30 degrees. Even, indoors the temperature is over 28 degrees. I had several batches of guppies dying due to HOT ambient. I now cover the top with black shade used to limit solar radiation to plants. Even the fish tank indoors gets ugly with algae) was once a day any time of the day was a discovery.

No fish died due to my irregular habits.

In fact fish started cleaning the tank of algae, unfortunately only a certain types of fish love algae.

Mind you even the guppies took to changing their habits.

What was the final outcome of my digital zeroing.

For the first time in 15 years, I got my experimental water garden into manageable business with exotic plants having a special display.

So digital zeroing can spring surprises and can be invigorating for your soul.

My final conclusion, it is better than practicing meditation and the plants will respond to your transformation with lovely flowers blooming on your foot path.

Who says plant cannot sense you.

They do better than fish.

Mind you pet fish are too domesticated to respond like he plants.

Guppies dying due to extremes of heat is a case in point.

Even the dog started practicing my tricks and did not seek any special attention.

It was accommodating and missing its meal time was not a big deal.

I had his treats in store as a precaution.

The bottom line is, the practicing digital zeroing is good for your soul and your loved ones.

This piece is at least two years OLD.

Why it is so painful to cancel an account in a Social Media

This time Hi5

I still wonder why social media are behaving like leeches. They are hell bent on increasing their share in the open but dirty market.

1. They start by offering you a free account.
That is the final bait.

In this consumerist world who will offer a free lunch?

2. Then they get every bit of your personnel data which even (probably) government agencies are not interested in.

The governments want to know you are security risk and or a xenophobic.

If you are an average criminal, they are not worried about.

In this country the previous as well as present regime want to be in book with the underworld in spite of clearing the so called terrorists.

They only want big fish.

In this country of ours, some government agencies actually promote average tricksters and they take them into their fold, get the dirty work done and dispose them unceremoniously like a pulp when it is a liability (we have a police ex-chief and a Sanganayaka involved in plundering archaeological artifacts) to the beholders.

Guys / Girls include those who steal our ballot paper for election rigging.

Some of them steal even the vote of old people in "Incarcerated Homes".

They (the criminals) are a wanted commodity come elections.

3. These social media know that the governments steal their data whenever and wherever possible.

Earlier it was done at the airport.

Now they scan you in your kitchen doing daily chores.

They do not have to be on the beat but remain in air conditioned cubicles.

What a luxury?

4. The government's paranoia for no reason to remain in absolute power has now infiltrated the social media.

They fear a bubble like in the credit crunch.

They fear the credit crisis.

Their profit margin that come free due to technology and that infiltrates into every nook and corner (luckily we have only 10% Internet penetration in spite of the massive drive) one does not have to sweat for one's tasks.

These guys also live in air conditioned living rooms and sit in front of the computer and do not have to travel like the ordinary guys and girls who toil, every day of the week.

They have cushy IT jobs.

The network of computers and various search engines do the talk.

They talk tall!

5. Invariably they become the worst pests of this century and of any country.

So they fear when a old guy like me want to opt out of their site for an ornamental coffin for the eternal peace six feet under.

They make de-registration, the most difficult job on this planet.

Why?

I do not know.

As a doctor, I can say it is an early sign of serious pathological entity not yet defined by the medical fraternity.

I coin the word Digital Mafia that metamorphoses to Digital Morphea

Getting out of Hi5 was as difficult as with FaceBook.

The details are hidden in the Full Site out of reach of the folks.

You have to contact them by email and tag with another entity and a token obtained for one to see the Full Site.

Even after getting there they will ask you various (FUD, Fear Uncertainty, and Doubt) silly questions.

Now they try to keep the data in their server and promote a reactivation scheme.

I don't think I can reactivate it from my coffin unless, I take a Tablet with Wireless connection to my final resting (rite's) site, the TOMB.

Rest in Piece Social Media.

I have raised this many moons ago.

What happen to my emails when I am DEAD and Disposed?

What happen/s to it / them when I am disposed or dispossessed?

Have I got to declare a Dying Dispossession?

In spite the technology they are unable to device a full proof method.

That is the case in point.

I tell old people please do not use Internet.

Visit your friends in person and enjoy a cup of Tea/Coffee or a pint of Larger.

That is more user friendly than social media.

Real people (not digital) need personnel touch.

I now practice digital zeroing when I am on holidays.

I really enjoy my privacy and holiday but enjoy every bit of flavor of food and drinks, I used to enjoy as a young guy (but going up in price for an average citizen).

I practice bit of Moment Meditation, too.

Life is to live.

Not to sit in front of an idiot box.

Creativity comes naturally, then.

See you next time.

I am trying to devise a method to opt out not log out.

Hope, I will have a Brain Wave when on digital holiday.

Asokaplus

www.ingramcontent.com/pod-product-compliance
Lightning Source LLC
Chambersburg PA
CBHW081249180526
45170CB00007B/2350